Thomas Moffett

**The Silkewormes and their Flies**

Thomas Moffett

**The Silkewormes and their Flies**

ISBN/EAN: 9783742831224

Manufactured in Europe, USA, Canada, Australia, Japa

Cover: Foto ©Klaus-Uwe Gerhardt /pixelio.de

Manufactured and distributed by brebook publishing software
(www.brebook.com)

Thomas Moffett

**The Silkewormes and their Flies**

# Silkewormes, and their Flies:

Liuely defcribed in verfe, by T. M.
*a Countrie Farmar, and an ap-*
prentice in Phyficke.

*For the great benefit and enriching of England.*

Printed at London by V. S. for Nicholas Ling, and
are to be fold at his fhop at the Weft ende of
Paules. 1 5 9 9.

# To the moſt renowned Patroneſſe, and noble Nurſe of Learning MARIE Counteſſe of Penbrooke.

Reat enuies Obiect, Worth & Wiſedoms pride,
Natures delight, Arcadia's heire moſt fitte,
Vouchſafe a while to lay thy taſke aſide,
    Let Petrarke ſleep, giue reſt to Sacred Writt,
Or bowe, or ſtring will breake, if euer tied,
Some little pawſe aideth the quickeſt witte:
    Nay, heau'ns themſelues (though keeping ſtil their way)
    Retrogradate, and make a kind of ſtay.

I neither ſing Achilles baneful ire,
Nor Man, nor Armes, nor Belly-brothers warres,
Nor Britaine broiles, nor citties drownd in fire,
Nor Hectors wounds, nor Diomedes skarres,
Ceaſe country Muſe ſo highly to aſpire:
Our Plaine beholds but cannot holde ſuch ſtarres:
    Ioue-loued wittes may write of what they will,
    But meaner Theams beſeeme a Farmers quill.

I ſing of little Wormes and tender Flies,
Creeping along, or basking on the ground,
Grac't once with thoſe thy heau'nly-humane eies,
Which neuer yet on meaneſt ſcholler fround:
And able are this worke to æterniſe,
From Eaſt to Weſt about this lower Round,
    Deigne thou but breathe a ſparke or little flame
    Of likeing to enlife for aye the ſame.

                Your H. euer moſt bounden.
                        T.    M.

# The Table.

# The Table.

FINIS.

# Faults escaped in Printing.

| page | line | | fault | | | correction |
|------|------|---|-------|---|---|------------|
| 5 | 11 | | the | | | thy |
| 7 | 3 | | euer | | | neuer |
| 7 | 14 | | courſer | | | Courſers |
| 9 | 19 | | priuate | | | Priuie |
| 17 | 3 | | hie | | | thie |
| 17 | 13 | | layes | | | laye |
| 19 | 2 | for | h harſes | reade | | herſes |
| 27 | 4 | | through | | | thorough |
| 29 | 10 | | through | | | thorough |
| 48 | 2 | | Enicthean | | | Erycthean |
| 56 | 1 | | us | | | as |
| 59 | 1 | | I any | | | If any |
| 66 | 15 | | dropt | | | drop |

# Of the Silke wormes and their Flies.

Ydneian Muse: if so thou yet remaine,
In brothers bowels, or in daughters breast,
Or art bequeath'd the *Lady of the plaine*,
Because for her thou art the fittest guest:
Whose worth to shew, no mortall can attaine,
Which with like worth is not himselfe possest:
  Come help me sing these flocks as white as milke,
  That make, and spinne, and die, and windle silke.

For sure I know thy knowledge doth perceiue,
What breth embreath'd these almost thingles things:
VVhat Artist taught their feete to spinne and weaue:
What workman made their slime a robe for kings,
How flies breed wormes, how wormes do flies con-
Frō natures womb, how such a nature springs, (ceiue:
  Whereof none can directly tell or reede,
  Whether were first, the flie, the worme, or seede.

A time there was (sweete heau'ns restore that time,)
When bodyes pure to spotlesse soules first knit,
Deuoyd of guilt, and ignorant of crime,
Vpright in conscience, and of harmelesse wit,
Disdaind to weare a garment nere so fine,
As deeming coates and couers most vnfit,
  Where nothing eie could see, or finger touch,
  Which God himselfe did not for good auouch.    Gen. I. verse 31.

B                    Yea

Yea, when all other creatures looked bafe,
As mindful onely of their earthly foode:
Or elfe as trembling to behold the place,
Where iudge eternall fate, and Angels ftood:
Then humane eyes beheld him face to face,
And cheekes vnftain'd with fumes of guiltie bloud,
    Defir'd no maske to hide their blufhing balles,
    But boldly gaz'd and pried on heau'nly walles.


The breaft which yet had hatcht no badde conceat,
Nor harbor'd ought in heart that God difpleaz'd,
Did it for filken waftcotes then intreate?
Sought it with Tyrian filks to be appeaz'd?
No, no, there was no neede of fuch a feate,
Where all was found, and members none difeaz'd:
    Nay more, The bafeft parts and feates of fhame,
    Were feemely then, and had a comely name.


*Gen.3.*   But when felfe-will and fubtile creepers guile,
Made man to luft, and tafte what God forbad,
Then feem'd we to our felues fo foule and vile,
That ftraight we wifht our bodies to be clad,
Seeing without, and in fuch great defile,
As reft our wittes, and made vs al fo mad:
    That we refembled melancholique hares,
    Or ftartling ftagges, whom euerie fhadow fcares.
                     Then

Then Bedlam-like to woods wee ranne apace,
Praying each tree to lend vs shade or leaues,
Wherewith to hide(if ought might hide) our face
From his al-seeing eyes,who al perceaues,
And with ful-brandisht sword pursues the chace,
Traitors of rest,of shade,and al bereaues:
    Permitting men with nothing to be clad,
    But shame,dispaire,guilt,feare,and horror sad.

These robes our parents first were deckt withal,
Then figtree sannes vppon their shame they wore:
Next, skinnes of beasts,(to shew their beastly fall)
Then,hairy cloathes,and wooll from Baa-lambs tore,
Last,Easterne wittes,from mane of Camels tall; *Plin.lib.11.ca.*
Made water-waued stuffe vnseene before, *10.& lib.24.*
   But til the floud had sinners swept away, *cap.12.*
   Nor Flaxe, nor Silke, did sinful man array.

For so it seeemed iust to Iustice eyen,
Defiled men to weare polluted things:
And Rebels not to clothe in Flaxe or line,
Which from the sacred loines of *Vesta* spring,
Cleane,knotlesse,straight,spotlesse,vpright,and fine,
VVhose floure is like fiue heau'nly-azurd wings,
   Whose slime is salue,whose seed is holsom food, *Plut.lib.de Isid*
   whose rinde is cloth,whose stuble seru's for wood *& Osir.*

4

*A most famous*
*spinner in Lydi-*
*a, of whom Ouid*
*6 metam.*

1

Or if 1 *Arachne* erst made sisters threed,
Was it thinke you, for euery man to weare?
Or onely for the sacrificers weede,
VVho of th'immortall priest a type did beare?
Wearing not aught that sprang from brutish seed,
But what from out it selfe the earth did reare:
    So that till holy priesthood first began,
    VVe neuer reade that linnen clothed man.

2

Yet some conceiue when 2 *Theban* singer wanne,
VVood-wandring wights to good and ciuill life,
(Which erst with beares and wolues in desarts ran,
Knowing no name of God, law, house, or wife).
That then his brother *Linus* first began
The Flaxmans craft (a secret then vnrise)
    Deuising beetles, hackels, wheeles, and frame,
    Wherwith to bruse, touse, spin & weaue the same.

But Silke (whereon my louing Muze now stands)
Was it the offspring of our shallow braine?
Spunne with these fingers foule? these filthy hands,
Tainted with bloud, reuenge, and wrongful gaine?
Ah no, who made and numbreth all the sands,
Wil teach vs soone that fancie to be vaine:
    Farre be it from our thoughts, that sinfull sence,
    Should make a thing of so great excellence.

Ne

Ne neede wee yet with 1 *Tuscane* Prelate flie,
To fictions strange, or wanton *Venus* eyen:
Who seeing *Pallas* taught from *Saturne* hie,
To clothe her selfe and hers with weaued line,
Yea all the Nimphs and Goddesses in skie,
To weare long stoles of Lawne and Cambrick fine:
    Fretted to see her selfe and boy new borne ,
    Left both to heau'n and earth an open scorne.

1
*Hieronimus vi-
da, Bishop of
Alba.lib.1.de
Bombyce.*

*Reuenge* she cri'de vnto the sire of *Ioue*,
As she lay hidde vnder th'Idalian tree:
Affoord some rayment from the house aboue,
If but to hide the shame of mine and mee.
So may thou learne from vs *The art of Loue*,
Whereby to winne each Ladies heart to thee.
    But grumbling Chuff reiected still her prayre,
    Whereat lamented heau'ns and weeping aire.

Then Cyprian Queene perceiuing that no cries
Could pierce the leaden eares of sullen Sire,
Straight lodg'd her sonne in faire 2 *Phillyraes* eies,
And caus'd him thence to darte vppe such a fire,
As had consum'd the very staires and skies,
Yea melted *Saturnes* wheeles with hot desire:
    Vnlesse that very houre he had come downe,
    And beg'd her aide, on whom he late did frowne.

2
*Oceanus his
daughter, a
most braue vir-
gin, Ouid 6 met.*

           B 3         How

How often,as his loue on *Pelion* hill
Stoopt downe to gather herbs for wounds and sores,
Strew'd he before her Tutsan,Balme,and Dill,
Long Plantaine,Hysope,Sage,and Comfrey moares?
Offring besides,the art and perfect skill,
Of healing bloudy wounds and festred coates:
   How oft (I say)did he eachiday descend,
    And bootelesse al his vowes and wooings spend?

He lou'd,she loath'd,he liked, she disdain'd:
He came,she turn'd,he prest, she ranne away,
Neither by words, nor gifts shee could be gain'd,
(For onely in her eies the Archer lay)
Regarding nought but (wherein she was train'd)
VVounds how to cure,and smartings to allay:
   As for the wound of Loue,she felt it none,
    And therefore litle heeded *Saturns* mone.

Thus thus perplext the chiefe and grauest God,
(Or rather God suppos'd of highest place)
Toucht now , nay throughly scourg'd with *Cupids*
Sent from the eyes but of a mortal face,    (rodde,
Flewe downe forthwith where *Venus* made abode,
And prostrate lying at her feete for grace:
   Promis'd the richest clothing for her Art,
    That now she did,or could desire in hart.

                            VVhe

VVho carelesse of reuenge, and innely grieu'd,
(True beauty aye is ful of rucful mone)
VVas euer wel til *Saturne* was releeu'd,
His inward griefes asswag'd, & sorrowes gone,
And finding him, of hope, and helpe, bereeu'd,
(For still *Phillira* was more hard then stone)
  Sith that, quoth she, the virgin scorns thy loue,
  Try whether craft and force wil make her moue.


Transforme thy selfe into a Courser braue,
(VVhat cannot loue transforme it selfe into?)
Feede in her walkes: and in a moment haue
VVhat thou hast woo'd to haue with much adooe:
VVhereto, consent the auncient Suter gaue,
In courser clothes, learning a maide to wooe,
  Filling ech wood with neighs and wihyes shrill,
  VVhilst he possest his loue against her will.


For lesson which, his Mistris to requite,
Not with vaine hopes in lieu of friendly deeds,
By *Maiae's* sonne (before it grew to night)
He sent a Napkin ful of little seeds,
Tane from the tree where *Thisbes* soule did light,
To make her selfe and boy farre brauer weeds,
  Than *Pallas* had, or any of the seu'n,
  Yea, then proud *Iuno* ware the Queene of heau'n.
              VVithal

1 *Mercurie, postmaster to Iupiter.*

Withall, by him he sent the mysterie
Of weauing silke, which he himselfe had found,
When chac'd from heau'n by sonnes owne trechery,
Hee was compel'd to wander here on ground,
Where, in the depth of griefe and pouertie,
The heigth and depth of Arts he first did sound:
  Yet would he this to none but her reueale,
  By whose deuise hee did *Phillyra* steale.


What? shall we thinke, that silke was a reward,
Bestow'd on craftie dame for aide vniust?
Would men, nay, ought they haue such hie regard,
Of that which was the lone and hire of lust?
Not so, what ere th'Italian Bishop dar'd
To faine for true, and giue it out with trust:
  Yet sith silke robes the blessed High-priest wore,
  They were not sure the first fruits of a whore.


*Plinius Secundus, lib. 11.cap. 2*

2
*Called Pamphilia, a most princely Damsell.*

*Vespasians* 1 Scribe affirmes in *Cean* Ile,
*Latous* 2 daughter, quicke of eye and wit,
Hunting abroad, times trauaile to beguile,
Chaunc'd at the length vnder a tree to sitte,
Where many silken bottoms hangd in piles,
One by another plac't in order fit.
  Shee tooke one downe, and with her faulcon eye,
  Found out the end that did the rest yntie.

                             Looke

Looke how the hungry Lambe doth friske and play,
With reftleffe taile, and head, and euery limbe,
When it hath met his mother gone aftray,
Who abfent blear'd and tear'd as much for him:
Or as *Aurora* leapes at breake of day,
Seeing her louely brother rife fo trim,
  No leffe that Princeffe triumph't (if not more)
  Finding out that which was not found before.


*Loues Schoolemafter* 1 records a tale moft fweete,     1 *Ouid lib.4.*
Of louers two that dwelt at *Babilon*,     *Metam.*
Equall of age, in worth and beautie meete,
Each of their fex the floure and paragon,
Next neighbours borne on fide of felfefame ftreete,
For twixt their parents houfes dwelled none,
  Him *Pyramus*, her *Thisbe* men did call,
  Coupled in heart, though feuered by a wall.


As neighbours children, oft they talke and view,
That neighbourfhip was formoft fteppe to loue,
Loue, which (like priuate plants) in fhort time grew,
Pales, wals, and eues, yea houfes and all aboue,
Nay Hymeneus feafts were like t'enfue,
And facred hands giue ring and wedding gloue,
  Had not vnhappie parents that forbad,
  Which to forbid, no caufe but wil, they had.

If louers fpake, it was now all by lookes,
None deign'd or durft be trouchman to their mind,
Paper was barr'd, and penne, and inke, and bookes,
Not any helpe thefe parted prifoners find,
But of a rift along the wal that crookes,
(A wall of flint, yet more then parents, kind)
   Which, were it old or new, none it efpies,
    But louers quicke, al-corner-fearching eyes,

This rift they vfde, not onely as a glaffe,
Wherein to fee daily each others face,
But eke through it their voyces hourely paffe,
In whifpring murmurs with a ftealing pace:
Sometimes when they no longer durft (alas) (place,
Send whifprings through, when keepers were in
   Yet would they fhift to blow through it a breath,
    Which fed & kept their hoping harts from death.

Enuious wal (fayd they) what wrong is this?
Why doth not loue or pittie make thee fal?
Or (if that be for vs too great a bliffe)
Why is thy rift fo narrow and fo fmall,
As to deny kind loue a kindly kiffe?
For which we neuer proue vnthankful fhal,
   Although in truth we owe inough to thee,
    Giuing our eyes and voyce a way fo free.

In

In vaine thus hauing plaind in place distinct,
When night approacht, they ech bad ech adew,
Kissing their wal apart where it was chinckt,
Whence louely blasts and breathings mainely flew:
But kisses staide on eithers side fast linckt,
Seal'd to the wal with lips and Louers glue:
  For though they were both thick and many cake,
  Yet thicker was the wal that did them breake.

Rose-fingred 1 Dame no sooner had put out
Nights twinckling fires and candles of the skie,
Nor *Phœbus* 2 brought his trampling steeds about,
Whose breath dries vp the teares of *Vestaes* 3 eie,
But swift and soft, without all noyse or showt,
To wonted place they hasten secretly,
  Where midst a many words muttred that day,
  Next midnights watch, each vowes to steale away.

*1 The morning, Homer. Iliad. 4.*
*2 The Sunne.*
*3 The earth.*

And lest when hauing house and cittie past,
They yet might erre in fields, and neuer meete,
At *Ninus* 4 tombe their *Rendes-vous* is plac't,
Vnder the Mulb'ry white, and hony-sweete:
Growing hard by a spring that ranne at waste,
With streames more swift then speedy 5 *Isters* feete.
  There they agreed in spite of spite to stand,
  Whē 6 Monarchs teame had past 7 *Bootes* hand.

*4 Which was without the gates of Babilon, towards the forrest Sabell. Enneiad. 1. cap. 6.*
*5 The swift riuer of Donawe.*
*6 The Charles waine.*
*7 The great star following Vrsa maior.*

C 2      Con-

Confent they did, and day confented too,
Whofe Coach ranne downe the feas in greater haft,
Then euer it was wont before to doo,
Loue-louing night approched eke fo faft,
That darkneffe leapt,ere twilight feem'd to go,
Wherat though fome gods frown'd,fome were
   Yet *Lethes* 1 brother did the louers keepe, (agaft,
   Chaining their guard with long and heauy fleep.

*1 Sleep the brother uf forgetfulneffe.Cic.lib. de nat. deorum.*

How feately then vnfparred fhe the doore?
How filent turn'd it on the charmed cheekes?
And being fcap't,how glad was fhe therefore?
How foone arriu'd where fhe her fellow feekes?
Loue made her bold, loue gaue her fwiftneffe more
Then vfually is found in weaker fexe,
   But all in vaine : nay rather to her ill,
   For hafte made wafte,and fpeede did fpeeding kil.

The grifly wife of brutifh monarch ftrong,
With new flaine prey,full panched to the chinne,
Foming out bloud,came ramping there along,
To filuer fpring,her thirft to drowne therein,
Whereat the fearefull maide in pofting flung,
(For 2 *Lucines* eye bewrayde the Empreffe grimme)
   Into a fecret caue : and flying,loft
   A fcarfe(for *Pyrams* fake)beloued moft.

*2 The Moonefhine.*

                       When

When sauage Queene had wel her thirst delayde,
In cooling streames, and quenched natures fire,
Returning to the place where late she prayde,
To eate the rest when hunger should require,
In peeces tore the scarfe of haplesse maide,
With bloudy teeth, and firie flaming ire,
  Whilst she (poore soule) in caue plaid least in sight,
  Fearing what should her loue befall that night.

Who comming later then by vow he should,
Perceiu'd a Lions footsteps in the sand,
Whereat with face most pale, and heart as cold,
With trembling feare tormented he doth stand.
But when he sawe her scarfe (wel knowne of old)
Embru'd with bloud, and cast on either hand?
  O what a sigh he fetcht? how deepe he gron'd?
  And thus, if thus : yea, thus he inly mon'd.

Shalt thou alone die matelesse, *Thisbe* mine?
Shall not one beast be butcher to vs both?
What? is my *Thisbe* reft of life and shine?
And shal not *Pyram* life and shining loath?
Mine is the cursed soule, the blest is thine,
Thou kep'st thy vow, I falsified mine oath,
  I came too late, thou cam'st (alas) too soone,
  Too dangerous standing, by a doubtfull moone.

O Lions fierce(or if ought fiercer be,
Amongst the heards of woody outlawes fell)
Rent,rent in twaine this thrise-accurfed me:
From out your paunch conuey my foule to hell:
Whofe murdring flouth,and not the fifters three,
Did *Thisbe* fweete,fweete *Thisbe* fowly quell:
     But cowards onely call & wifh for death,
     Whilft valiant hearts in filence banifh breath.


Then ftooping,ftraight he took his fcarfe fró ground,
And bare it with him to th'appoynted place,
Kifsing it oft,watring each rent and wound,
With thoufand teares,that trailing ranne apace.
Salt teares they were,fent from his eyes vnfound,
Yea falter then the fweate of Oceans face:
     At laft (hauing vnfheath'd his fatall blade)
     Thus gan he cry,as life beganne to fade.


Hold earth, receiue a draught eke of my bloud,
(And therewith lean'd vppon his fword amaine)
Then falling backward from the crimfin floud,
Which fpowted forth with fuch a noyfe and ftraine,
As water doth,when pipes of lead or wood,
Are goog'd with punch,or cheefill flit in twaine,
     Whiftling in th'ayre, & breaking it with blowes,
     Whilft heauie moyfture vpward forced flowes.
                          The

The Mulb'ry ſtrait(whoſe fruit was erſt as white
As whiteſt Lilly in the fruitfullſt field)
Was then and euer ſince in purple dight,
Yea euen the roote no other ſtaine doth yeeld,
With blackiſh gore being watred all that night,
In morneful ſort, which round about it wheel'd,
   Onely her leaues retaind their former hue,
   As nothing toucht with death of louer true.


No ſooner was hee falne, and falling, freed
Of perfit ſence: but ſhe ſcarce rid of feare,
Returnes againe to ſtanding fore agreed,
Not dreaming that her loue in kenning were,
Her feete, her eyes, her heart and tongue made ſpeed,
To vtter all things lately hapned there,
   And how ſhe ſcap't the Lioneſſes clawes,
   By letting fall a ſcarfe to make her pawſe.


But when ſhe vewd the newly-purpled face
Of Berries white: that changing chang'd her mind,
New ſignes perſwade her, that is not the place,
By either part to meete in fore aſsign'd.
Thus doubting whilſt ſhe ſtood a little ſpace,
She heard a fluttering carried with the winde,
   And viewed ſomewhat ſhake in quiu'ring wiſe,
   Which ſtraite reuok't hir feete, but more her eies.
                            Her

Her lippes grew then more pale then paleſt Boxe,
Her cheekes reſembled Aſhwood newly feld,
Grayneſſe ſurpriz'd her yellow amber locks,
Not any part their liuely luſtre held:
Yea euen her vent'rous heart but faintly knocks,
Now vp, now downe, now falne, now vainly ſweld,
    Toſt like a ſhippe when 1 *Corus* rageth moſt,
    That ankers hath, and maſts and maſter loſt.

1 One of the Northweſt windes.

But when ſhe knew her faithfull fellow ſlaine,
O how ſhe ſhrikt and bruz'd her guiltleſſe arme,
Tearing her haire, renting her cheekes in vaine,
On outward parts, reuenging inward harmes,
Making of teares and bloud a mingled raine,
Wherwith ſhe *Pyram* drencht, & then thus charmes:
    Speake loue, O ſpeake, how hapned this to thee?
    Part, halfe, yea all of this my ſoule and mee.

Sweete loue, reply, it is thy *Thisbe* deare,
She cries, O heare, ſhe ſpeakes, O anſwere make:
Rowſe vp thy ſprights: thoſe heauie lookers cheere,
At which ſweete name hee ſeemed halfe awake,
And eyes with death oppreſt, againe to cleere.
He eyes her once, and eying leaue doth take,
    Euen as faire *Bellis* 2 winkes but once for all,
    When winters 3 vſher haſtneth ſummers fall.

2 The white Daiſy.
3 Harueſt.

                       When

When afterwards she found her scarfe al rent,
His iu'ory sheath voide eke of rapier gilt:
And hath his hand (quoth she) thy soule hence sent?
And was this bloud by this thy rapier spilt?
Vnhappy I:but I no more lament,
But follow thee euen to the vtmost hilt.
   **I was** the cause of al thy hurt and crosse,
   **Hold**, take me eke a partner of thy losse.


Whom onely death could from me take away,
Shal death him take from me against my will?
Not so,his power cannot *Thisbe* staye:
Who euen in death wil follow *Pyram* still,
His blade (yet warme) then to her brest she lays,
And falne thereon thus cri'de with crying shrill:
   Parents vniust which vs deny'd one bed,
   Enuy vs not one toomb when we be dead.


And al you heau'nly hostes allot the same:
And thou O tree,which coueret now but one
(One too too hot, for 1 so imports his name)
But couer shalt two carcasses anone:
Weare signes of bloud from both our harts that came
In mourning weed our mischiefes euer mone.
   She dead:Tree,Sires,& Gods gaue what she praide,
   Black growes the fruit, and they together laide.

1 *Pyramus signifieth as much as fiery.*

     D         Since

*1 Natal.Com.*
*lib.vlt.Mytho.*
Since which time eke some other (1) Authors faine,
Their humming soules about these haplesse trees,
To be transported from th'Elysian plaine,
Into the snowy milke-white Butterflyes:
Whose seedes when life and moouing they obtain,
How e're they spare the fruit of Mulberies,
    Leaue yet no leaues vntorne that may be seene,
    Because they onely still continude greene.


Yet that there might remaine some *Pyramis*,
And euerlasting shrine of *Pyrams* loue,
When leaues are gone, and summer waining is,
The little creepers neuer cease to moue,
But day and night (placing in toyle their blisse)
Spinne silke this tree beneath and eke aboue:
*2 Egge-like.*
    Leauing their ouall (2) bottoms there behind,
    To shewe the state of eu'ry Louers mind.


For as in forme they are not wholly round,
As is the perfit figure of the skie,
So perfit loue in mortals is not found,
Some little warts or wants in all we spie,
Nay eu'n as fine and course silke there abound,
The best beneath, the worst rold vp more hie,
    So sometimes lust o're-lieth honest loue,
    Happy the hand that keepes it from aboue.

A·

Againe, as thefe fine troupes themfelues deuoure,
Spinning but filken hharfes for their death:
VVhich done,they dye therein,(by Natures power
Transform'd to flies that fcarce draw one months
So louers fweet is mingled ftil with fower, (breath)
Such happe aboue proceeds or vnderneath,
    That ftill we make our loue our winding fheete,
    VVhilft more we loue,or hotter then is meete.

Others(1)report,there was and doth remaine
A neighbour(2)people to the *Scythian* tall,
Twixt *Taurus* mount and *Tabis* fruitful plaine,
Moft iuft of life,of fare and diet, fmal,
Louers of peace,haters of ftrife and gaine,
Graye ey'd,redde cheek'd,and amber-headed all,
    Refembling rather Gods then humane race,
    Such grace appeard in words,in deeds,and face.

1 *Plin.lib.6.*
*cap.*17.
2.*Called Seres.*

VVhofe righteous life and iuftice to requite,
(Whether with wind or raine,no man doth know)
God fent vnto them filke-wormes infinite,
In Aprils wane when buds the mulb'ry flow,
Which here and there in euery corner light,
With fixe white feete and body like to fnow:
    Eating each leafe of that renowned tree,
    The matter of thefe filken webbes we fee.
              D 2       Thefe

These webbs for wares they on their coast exchange:
For alien none must come into the Land,
T'infect their people with religions strange,
And file their temples with polluted hand:
Neither do they to other nations range,
New fashions, rites or manners t'understand:
    Better they haue at home, where euery slaue
    Weares silks as rich as here our Princes braue.


These be the tales that Poetizers sing,
Of Silken-worme, and of their seed and meate:
*1 VVherof only* Sweete, I confesse, and drawn from 1 Helique spring,
*the muses drank,* Full of delighting change, and learning greate.
*as Poets imagine.* Yet, yet, my Muse dreames of another thing,
And listeth not of fictions to entreate.
    Saye then (my Ioye) say then, and shortly reede,
    whē silk was made, & how these silkworms breed.


Was it think'st thou found out by industry?
Inspir'd by vision or some Angells word,
*2 Melchisedec.* When first the name of sacred Maiesty,
Was giuen from heau'n to 2 *Salems* priest and Lord?
Did not before tenne thousand Silk-worms lye,
And hang on euery tree their little cord?
    Yes, but (like *Hebrues* harps on *Babels* plaine)
    Vntoucht and vse-lesse there it hang'd in vaine.

                                 Before,

Before, most men liu'd, either naked quite,
Or coursly clad in some beasts skinne or hide:
The best were but in linnen garments dight,
Wherein themselues the greatest men did pride:
Yea afterward in time of greatest light,  *Mat. 1ƒ.*
When chiefe Baptizer preach't in desart wide,
   Where said he, silken robes were to be sought,
   But in kings courts ? for whome they first were
                        (wrought.


Though whether worme or flye were formed first,
No man so right can tel as wrong presume:
Yet this I hold. Till all things were accurst,
Nothing was borne it selfe for to consume.
No Caterpillers then which venture durst,
To rauish leaues, or tender buddes to plume:
   For onely life and beauty liu'd in trees,
   Til falling man caus'd them their leaues to leese.


The earthly heards and winged posts of skye,
And eu'ry thing that mou'd on Eden ground,
Fed first on hearbs (as Duke of *Horeb* hie,  *1 Moses.*
Author of Natures story most profound,
Sets downe to vs for perfit verity,
(Gaines aide of none but fooles and wittes vnsound)  *Gen.2.verse 29*
   When for mans foode trees eke allotted were,
   Which from themselues did fruit or berries beare.

Durst

Durst then the finest worme but touch the meate,
Or dish which for his soueraigne was ordain'd?
Durst they figges, nuts, peares, plummes or mulb'ries
Before their lord with treasō foule was stain'd? (eate
No certs no, but when ambitious heate,
Reuok't the blisse which sinnelesse Sire had gain'd:
　　Then wormes in common fed with vs, and tore
　　Our trees, our fruits, yea eu'n our selues therefore.


1 Herod.
Act.12.
2 Antiochus E-
piphanes.
3 Plato, who di-
ed eaten of lice,
as Diogenes La-
ertius writeth.
Say Romanes, heau'nly-humane (1) Orator,
Whose words dropt sweeter then *Hymettus* dewe:
Say (2) *Salems* scourge and *Iudaes* tormentor,
Whose very name doth pomp and glory shewe:
Say 3 thou whose writtes men as diuine adore,
Inspir'd from heau'n with knowledge giuen to few?
　　What are you now? what liuing were you then
　　But worms repast, though wise and mighty men?


Foule-footed bird, that neuer sleepest well
Nor fully, but on highest pearch do'st breathe:
Whose outward shreeks bewray an inward hell,
Whose glistring plumes are but a painted sheathe:
Whose taile, though it with pride so lofty swel,
Yet hides it not thy blacknesse vnderneath.
　　Tell me: what hast thou got by climing thus,
　　But to thy selfe a shame, and losse to vs?

To

To vs alone?nay ftowteft Okes likewife,
Hard-hatted willowes by the water fide,
Sweete Cedar wood which fome thinke neuer dies,
And 1 Daphnes tree though greene in winters tide, *1 The Bay.*
Yea ftone, and fteele,and things of higheft prize,
From natures womb that flow in greateft pride:
  What are they al but meate for wormes and ruft?
  Two due reuengers of ambitious luft.

Before thou waft, were Timber-worms in price,
And fold for equal weight of pureft gold?
Fed 3 creeping birds one barke-deuouring lice?
Were filk-worms from 4 *Serinda* brought and fold?
Deuoured they the leaues of tree moft 5 wife,
With fury fuch as now we do behold?
  Rather beleeue as yet they were not borne,
  Or onely fed on graffe,on hearbs, or corne.

*2 Called Coffi, which being fat, were counted a moft daintie dish in Rome. Cal.Sec.lib. 28. An.lect.*
*3 Titmife.*
*4 The firft and primit all place whence they were brought into Europe. Polyd.virg. lib. 11.de inuent. &c.*
*5 The Mulbery is called the wifeft tree,becaufe it neuer buddeth till all danger of cold be gone.*

For fith their chiefeft vfe is to arraye
This little breathing duft when time requires,
VVith gallant guards and broydred garments gaye,
VVith fcarfs,vales, hoodes,and other foft attires:
VVhofe fenfe from fenfe is fled fo farre away?
Whofe mind to beare fo wrong a thought confpires,
  As once to deeme thefe Silken-mercers fent,
  VVhen nakedneffe was mans chiefe ornament?
                                                But

But sith they are,and therefore framed were,
Which first was fram'd?the egge?the worme? or flie?
No doubt the flie,as plainely shall appeare,
To all that haue but an indiff'rent eye,                (beare,
Though twoo 1 great Clarks contrary thoughts did
And sentence gaue,without iust reason why,
   That egges were made before the hardie Cocke
   Beganne ro ttead,or brooding henne to clocke.

*1 Euangelus in Macrobius lib.4. sat.cap.3.& Firmus in Plutarch.lib.2.symp quest.3.*

Pretend they did,that least and simplest things,
(Which none train'd vp in reasons schoole gainsay)
Of things compounded are the formost springs,
Eu'n as a lumpe of rude and shapelesse clay,
Into the mould a Moulder cunning brings,
And by degrees compels it to obey:
   Forming by art what he in mind fore-thought,
   Out of a masse that iust resembled nought.

So eke though egges seeme things confused quite,
And farre vnlike what afterwards they prooue:
Yet formost place they challenge by their right,
For who e're saw a cock or henne to mooue,
Till first they came from out the yolke and white,
And time,and heate,and place,and sitters loue,
   Had formed out a nature from the same,
   Deseruing wel anothers natures name?

Springs

Springs not from egges that huge 1 Leuiathan,
The Tortesse eke, and bloudy Crocodile?
Fiſh, Lyzards, Snakes, and 2 Skippers African,
VVhoſe hurtful armies waſte the coaſts of Nile?
Nay if with one fitte word the world we ſcanne,
May it obtaine a fitter name or ſtile,
　Then ahat we ſhould a common egge it call,
　VVhich giueth life and forme and ſtuffe to all?

1 The VVhale

2 Locuſts or graſhoppers.

Nay, did not once that cheerefull brooding ſp'rite,
Before the earth receiued forme or place,
Sitte cloſely like a henne both warme and light,
Vpon the wauing neſt of mingled maſſe,
VVhilſt yet nights torches had obtain'd no light
Nor Sunne as yet in circled rounds did paſſe?
　Yes, yes: the words are ſo apparant plaine,
　That to deny them, were but labour vaine.

Gen. 1, verſe 2.

Theſe ſome do vſe with other arguments,
To proue that ſeede and egges were firſt in time.
VVreſted from quires of ſacred Teſtaments,
And thoſe of heathen wittes the chiefe and prime:
VVhich for authentique held by long deſcents,
If I gaineſay, perhaps may ſeeme a crime:
　Yet rather would I carry crime and ſcorne,
　Then falſely thinke, imperfect things firſt borne.

For reason saith, and sense doth almost sweare,
Natures entire to be created furst:
Bodies t'haue beene before the members were,
The sound before the sicke, the whole, the burst,
That confidence had time when lacked feare,
That blessed state fore-went the state accurst:
   Briefely, al bodyes that begotten beene,
   **Were** not before created bodies seene.

Now what are seedes and egges of wormes or foule,
But recrements of preexisting things,
The bodies burden voyd of life and soule?
Yea, from themselues corruption onely springs,
Vnlesse by brooders heate (as from the whole)
They changed be to belly, feete, or wings:
   Resembling them now metamorphosed,
   In, by, and from whose essence they were bred.

*Dibilus and Sentcio, their arguments against Firmus and Euangelus, of whom at large in Macrobius and Plutarke.*

Yea, vsual phrase such dreames confuteth quite,
For neuer man, *this is an egges henne* sayd,
But *this a hennes egge is*, shewing aright,
That egges are things by former natures layde,
Begotte of mingled seede by day or night,
Neither with skinne, nor shell, nor forme arrayd,
   Till long they haue abode in natures nest,
   And wearied womb be with their weight opprest

A-

Againe, to thinke that feede was made before,
The fubftance whence it is ingendered,
(Namely from out much nutrimental ftore,
Through exceffe of humours perfited)
Or elfe to gheffe it formed was of yore,
Ere pipes were laid through which it fhould be fhed,
   What is it but to dreame of day or night,
   E're darkneffe were, or any fhew of light?

Sith eke all winged creatures by one day,
Are elder then the heards that crawle and creepe,
Conclude with truth and confidence wee may,
All flies were made ere wormes beganne to peepe,
Both they which all day long at bafe do play,
And night once come, do nothing elfe but fleepe,
   And thefe which onely liue to leaue a feede,
   From whence the neuer-idle fpinfters breede.

*Gen. I. verfe 20 & 24.*

Silke-flies I meane, which not one breaft alone,
But all throughout, on head, wings, fides, and feete,
Befides pure white, elfe colour carry none,
For creatures pure, a colour thought moft meete,
Martial'd the firft of all in glorious throne,
Whereon fhall fit the Lord and Sauiour fweete,
   Who with tenne thoufand Angels all in white,
   Shal one day iudge the world with doom vpright

No spotte on them, as els on eu'ry flye,
Bycause in them no follies euer grew,
No crimson redde doth for reuengement crye,
No wauering watchet, where al harts be true:
No yellow, where there is no Iealousie:
No labour lost, and therefore voide of blue:
    No peachy marke to signifie disdaine,
    No greene to shew a wanton mind and vaine.


No orenge colour, where there wants despight,
No tawny sadde, where none forsaken be:
No murry, where they couet nought but light,
No mourning black, where al reioyce with glee:
In briefe, within, without, they are al white,
Wearing alone the badge of chastity:
    Bycause they onely keepe themselues to one,
    Who being dead, another chuse they none.


True Turtles mine, begotten with the breath,
Not of a lewd lasciuious mortal *Ioue*:              (death,
Whose lawe was lust, whose life was worse then
VVhose incests did defile both wood and groue,
But with the breath of him who vnderneath
Rules *Stigian* king, and heau'nly hosts aboue,
    Assist me if I erre in setting forth
    Your birth dayes story, and surpassing worth .
                                        Assoone

Assoone as light obtain'd a fixed seate,
(which equally was first spread ouer all,
Giuing alike, both glistring, shine, and heate,
To euery place of this inferiour ball)
Two master-lamps appear'd in welkin great,
Th'one king of day, whom Poets *Phœbus* call,
    And th'other *Phœbe*, soueraigne of the night,
      Twinnes at one instant bred and borne of light.

*Genesis 1.*

Him heau'nly Martiall high, in Pallace plac't,
Built all of cleere and through-shining gold,
With columnes chrysolite most brauely grac't,
And flaming rubies, glorious to behold,
Wearing about his yellow-amber wast,
A sloping belt, with studs twise six times told,
    Wherein were grau'n most artificially,
      Twelue stately 1 Peeres of curious imagery.

*1 The twelue signes in the zodiake.*

About him, as in royall Coach hee sate,
Attended Houre, Day, Minute, Month, and yeare,
Spring, Summer, Haruest, Winter, Morning, Fate,
With Instancie, who then was driuer there,
Whipping his fiery steedes from 2 *Libraes* gate,
Not suffring them to stand still any where,
    Saue once in *Gibeon* when fiue kings were slaine,
      By first-made 3 Champió with their faithles train.

*2 For it was then ful haruest and not spring-time, as the vulgarsort do hold. 3 Iosua cap. 10.*

His sisters court built al of siluer tri'de,
And Iu'ory charret set with Diamons,
Embost with Orient pearles on either side,
Wheeld al with Saphires, shod with Onyx stones,
Declar'd in what great pompe she first did ride
Amongst the other twinckling Paragons,
  Before her honour suffred an eclipse,
    Through serpents guile, and womans greedy lips.

Her handmaids then were perpetuity,
Constant proceeding, and continuance:
No shew of change or mutability
Could iustly then themselues in her aduance:
Her face was ful and faire continually
Not altering once her shape or countenance, (made,
  Till those lights chang'd for whom al lights were
    And with whose fall the heau'ns began to fade.

*1 Oceanus is the*
*king, & his wife*     Yet still on her wait (1) *Ocean* and his wife,
*Thetis is coun-*      *Nais* (2) the faire, and al the watry crue,
*ted the Queene*       Nights, Riuers, Flouds, Springs, hauing else no strife,
*of the seas.*         Then who may formost proffer seruice due:
*2 The Lady of*        Bloud, choller, phlegme, (the rootes and sappe of life)
*the riuers.*          Are at her beck, waining or springing new,
                         According as from throne celestiall,
                           She deignes to shine in measure great or small.
                                                                      When

When they were crowned now in royall thrones,
And entred in their first and happiest race,
Amongst those glistring pointed Diamons,
Which cut out times proportion, lotte, and space:
Behold the earth with heauy burden grones,
And praies them both to eie and rue her case:
　　And with their friendly hands and meeding art,
　　To hasten that which ready was to part.

For eu'n next morne the *All-creating Sire*
Had sent abroad, I know not I, what word:
Much like to this, *Let Sea and earth conspire*
*All winged troupes the world for to afford:*
Wherewith the aire: euen to the desart fire,
Was so with great and little flyers stor'd,
　　That none but winged people sawe the eies,
　　Of any star or planet in the skies.

*Gen.1.*

*1 So called by*
*Pyndarus, be-*
*cause nothing*
*liues in it.*

O how it ioyes my hart and soule to thinke
Vpon the blessed state of that same daye?
When at a word, a nodde, yea at a winke,
At once flew out these winged gallants gay,
Tide each to each in such a friendly linke,
That eu'n the least did with the greatest playe:
　　The doue with hawks, the chickens with the kite.
　　Fearelesse of wrong, rage, cruelty, or spite.

Pert

Pert marlins then no grudge to larkes did beare,
Fierce goshawkes with the Phesants had no warre,
Rau'ns did not then the Eagles talens feare,
Twixt Cuckoes and the Titlings was no iarre,
But coasted one another eu'ry where
In friendly sort, as louers woonted were:
   For loue alone rul'd all in eu'ry kind,
   As though all were of one and selfe same mind.

How safely then did these my Turtle-soules
Disport themselues in *Phœbus* cheerefull shine?
How boldly flew they by the iayes and owles,
Dreadlesse of crooked beakes or fiery eyen?
Nay, who in all the flocks of winged soules
Said once in heart, This pris'oner shal be mine?
   When none as yet made other warre or strife,
   Then such as 1 *Hymen* makes twixt man & wife.

1 *A Poeticall God, and suppo-sed instructor of brides and bride-groomes.*

But since the fall of parents pufft with pride,
Not onely men were stainde in viciousnesse,
But birdes, and beasts, and wormes, and flies beside,
Declining from their former petsitnesse,
Did by degrees to imperfections slide,
Tainted with pride, wrath, enuie, and excesse:
   Yea, then the husband of one onely henne,
   Was afterwards contented scarse with tenne.

                      Hence

Hence, gowts in cocks, and swelling paines appeare,
Hence, Partridge loynes so feeble we do view,
Hence, sparrow treaders liue out scarce a yeare,
Hence, leprosie the Cuckoes ouergrew:
Breefely, none did in true loue perseuere:
But these white Butterflies and Turtles true,
　　Who both in life and death do ne're forsake
　　Her, whom they once espoused for their make.


They choose not (like to other birds and beasts)
This yeare one wife, another wife the next,
Their choyse is certaine, and still certaine rests,
With former loues their mindes are not perplext,
Hee yeeldes to her, she yeelds to his requests,
Neither with feare nor ielosie is vext:
　　She clippeth him, hee clippeth her againe,
　　Equall their ioy, and equall is their paine.


Remember this you fickle hearted Sires,
Whom lust transporteth from your peereles Dames,
To scorch your selues at foule and forraine fires,
Wasting your health and wealth in filthie games,
Learne hence (I say) to bridle badde desires,
Quenching in time your hot and furious flames,
　　Let little flies teach great men to be iust,
　　And not to yeeld braue mindes a prey to lust.
　　　　　　　F　　　　　When

When thus they were created the first day,
Alike in bignesse, feature, forme and age,
Cladde both alike in soft and white array,
And set vppon this vniuersall stage,
Their seuerall parts and seates thereon to play,
Amidst the rest of natures equipage:        (thought)
    Who then suppos'd (as since some fooles haue
    That little things were made & seru'd for nought.


Diswitted dolts that huge things wonder at,
And to your cost coast daily ile from ile,
To see a Norway whale, or Libian cat,
A Carry-castle or a Crocodile,

*3 Heraclitus,*
*that euer wept.*
*2 Democritus*
*that euer laugh*
*ed as the worlds*
*folly.*

If leane Ephesian (1) or (2) th'Abderian fat
Liu'd now, and saw your madnesse but a while,
    What streaming flouds would gush out of theyr
    To see great wittols little things despise? (eies,


When looke, as costliest spice is in small bagges,
And little springs do send foorth cleerest flouds,

*3 Called Onis*
*in English.*

And sweetest (3) *Iris* beareth shortest flagges,
And weakest *Osiers* bind vp mighty woods,
And greatest hearts make euer smallest bragges,
And little caskets hold our richest goods:
    So both in Art and Nature tis most cleere,
    That greatest worths in smallest things appeare.
                                                    What

What wise man euer did so much admire
*Neroes* (1) Colossus fiue score cubits hie,
As *Theodorus* Image cast with fire,
Holding his file in right hand hansomly,
In left his paire of compasses and squire,
With horses, Coach, and footmen running by
    So liuely made, that one might see them all?
  Yet was the whole worke than a flie more small.

*1 Made by Zo-nodorus: of which, and alse of Theodorus image, more in Plin. lib. 34. cap. 7. & 8.*

Nay, for to speake of things more late and rife,
Who will not more admire those famous Fleas,
Made so by art, that art imparted life,
Making them skippe, and on mens hands to seaze,
And let out bloud with taper-poynted knife,
Which from a secret sheathe ranne out with ease:
    Thē those great coches which thēselues did driue,
    With bended scrues, like things that were aliue?

*2 Made by Gawen Smith. Anno. 1586.*

Ingenious (3) Germane, how didst thou conuey
Thy Springs, thy Scrues, thy rowells, and thy flie?
Thy cogs, thy wardes, thy laths, how didst thou lay?
How did thy hand each peece to other tie?
O that this age enioy'd thee but one day,
To shew thy Fleas to faithlesse gazers eye!
    That great admirers might both say and see,
    In smallest things that greatest wonders bee.

*3 Ioannes Re-giomontanus: of whom Ramus at large in Proem. lib 2. Math.*

Great was that proud and feared Philiftine,
Whofe launces fhaft was like a weauers beame,
VVhofe helmet, target, bootes, and brigandine,
VVeare weight (1)fufficient for a fturdy teame,
VVhofe frowning lookes and hart-difmaying eyne,
Daunted the tallest king of *Ifraels* realme:
Yet little shepheard with a pibble stone,
Confounded foone that huge and mighty one.

*1 For they weied 6000 Shekles of braffe.*

Huge fiery Dragons, Lions fierce and strong
Did they fuch feare on cruel (2)Tyrant bring,
VVith bloudy teeth or tailes and talens long,
VVith gaping Iawes or double forked fting,
As when the fmalleft creepers ganne to throng,
And feize on euery quicke and liuing thing?
No, no. The Egyptians neuer (3)feared mice,
As then they feared little crawling lice.

*2 Pharaoh.*

*3 Yet for fome of them they honoured their Gods in the forme of cats. Plaut. lib. de If. & ofi.*

*4 A moft famous trumpeter. Plin. lib. cap. 56.*

Did euer (4)Pifeus found his trumpet fhrill
So long and cleere, as doth the summer Gnat,
Her little cornet which our eares doth fill,
Awaking eu'n the drowzieft drone thereat?
Did euer thing do *Cupid* fo much ill,
As once a(5)Bee which on his hand did fquat?
Confeffe we then in fmall things vertue moft,
Gayning in worth what they in greatneffe loft.

*5 /nareten in one of his latter Odes.*

But

But holla, Muſe,extol not ſo the vale,
That it contemne great hilles,and greater skie,
Thinke that in goodneſſe nothing can be ſmall,
For ſmalneſſe is but an infirmitie,
Natures defect,and otſpring of ſome fall,
The ſcorne of men, and badge of infamy?
    For ſtill had men continued tall and great,
    If they in goodneſſe ſtill had kept their ſeate.


A little diſmall fire whole townes hath burnd,
A little winde doth ſpread that diſmall fire,
A little ſtone a carte hath ouerturnde,
A little weede hath learned to aſpire,
The little Ants(in ſcorne ſo often ſpurnd)
Haue galles : and flies haue ſeates of fixed ire.
    Small Indian gnattes haue ſharpe and cruel ſtings,
    Which good to none,but hurt to many brings.


And truely for my part I liſt not prayſe
Theſe ſilke-worme-parents for their little ſiſe,
But for thoſe louely great reſplendant rayes,
Which from their woorks and worthie actions riſe,
Each deede deſeruing well a Crowne of bayes,
Yea,to be grauen in wood that neuer dies:
    For let vs now recount their actions all,
    And truth wil proue their vertues are not ſmall.

F 3            Firſt

First, though fiue Males be brought to Females ten,
Yet of them al they neuer chuse but fiue,
Each takes and treads his first embraced henne,
With her he keepes, and neuer parts aliue:
And when he is enclos'd in Stygian penne,
Desireth she one moment to sutuiue?
    No, no, but strait (like a most louing bride)
    Flies, lies, and dies, hard by her husbands side.

Anno. Dom.
1579. when I
was in Italy.

In Tuscane towres what armies did I view
One haruest, of these faithful husbands dead?
Bleede, O my heart, whilst I record anew,
How wiues lay by them, beating, now their head,
Sometimes their feet, and wings, & breast most true,
Striuing no lesse to be deliuered,
    Then *Thisbe* did from vndesired life,
    When she beheld her *Pyram* slaine with knife.

But whilst they liue, what is their chiefest worke?
To spinne as spiders do a fruitlesse threed?
Or Adder-like in hollow caues to lurke,
Till they haue got a curst and cankred seed?   (fork,
(Whose yong ones therfore, with dame Natures
Iustly gnaw out the wombs that did them breed:)
    Or striue they Lion-like to seize and pray,
    On neighbours herds or herds-men by the way?
                     Delight

Delight they with ſtrange 1 Ants & Griphins ſtrong,
To hoord vp gold and eu'ry gaineful thing?
Liue they not beaſts, and birds, and men among,
Committing nought that may them damage bring?
O had I that fiue-thouſand-verſed ſong,
Which(2)Poet prowd did once with glory ſing,
 That whilſt I write of theſe ſame creatures bleſt,
 In proper words their worth might be expreſt,

*1 Of whō Pli-
ny writeth, lib.
11.cap.31.*

*2 Thamyris,
who wrote
5000. verſes of
the worlds crea-
tion Zetzes,7.
chil.hiſtor.108*

What wil you more? they feede on nought but aire,
As doth that famous bird of Paradice,
They liue not long, leſt goodneſſe ſhould empaire,
Or rather through that(3) Hagges enuious eyes,
That ſits,and ſitting, cuts in fatall chaire
That threed firſt off,which faireſt doth ariſe:
 Affording crowes and kites a longer line,
 Then fliers ful of gifts and grace diuine.

*3 Atropos.*

When maker ſaid to eu'ry bodied ſoule,
*Encreaſe,encreaſe,and multiply your kinde:*
What he or ſhe of al the winged ſoule
So much fulfill'd their ſoueraigne-Makers minde,
As theſe two flies? who coupled three dayes whole,
Left on the fourth more ſeeds or egges behind
 Then any bird:yea then the fruiteful wrenne,
 Numbred by tale a(4)hundred more then tenne.
       **Which**

*Gen.1.*

*4 Sometimes,
more,ſeldome
fewer.*

Which donne,both die,and die with cheerefull hart
Bycause they had done al they bidden were,
Might we from hence with conscience like depart,
How deare were death? how sweet & voyd of feare?
How little should we at his arrowes start?
If we in hands a quittance such could beare
　　Before that iudge, who looks for better deedes,
　　From men then flies, that spring of baser seeds.

*Ψυχὴ is all one name in Greeke for a soule and a butterflie.*

Go worthy soules (so (1) witty *Greeks* you name)
Possesse for aye the faire *Elisian* greene:
Sport there your selues ech Lording with his Dame,
Enioy the blisse by sinners neuer seene:
You liu'd in honour, and stil liue in fame,
More happy there,then here is many a Queene:
　　As for your seeds committed to my charge,
　　Take you no care : I'le sing their worth at large.

*2 The Lady of the plaine.
3 Miraes daughter.
4.5.6.7.8. Gentlewomen attending vpon Mira and her daughter.*

Weepe not faire(2)*Mira* for this funeral.
Weepe not(3)*Panclea,Miraes* chiefe delight,
Weepe not(4)*Phileta*,nor(5)*Erato* tall:
Weepe not(6)*Euphemia*, nor(7)*Felicia* white:
Weepe not sweete (8)*Fausta*:I assure you all,
Your cattels parents are not dead outright:
　　Keepe warme their egges,and you shall see anone,
　　From eithers loynes a hundred rise for one,
　　　　　　FINIS.

## ꙮ The second booke of the Silke-
### Wormes and their Flies.

O Thou whose sweet & heau'nly-tuned Psalmes
The heau'ns theselues are scarce inough to praise!
Whose penne diuine and consecrated palmes,
From wronging verse did *Royall Singer* raise,
Vouchsafe from brothers ghost no niggards almes,
Now to enrich my high aspiring layes,
   Striuing to ghesse, or rather truely reede,
   What shall become of all this little breede.


This little breede? nay euen the least of all,
The least? nay greater then the greatest are:
For though in shew their substance be but small,
Yet with their worth what great ones may compare?
What egges as these, are so much sphericall
Of all that euer winged Natures bare?
   As though they onely had deseru'd to haue,
   The selfe same forme which God to heauens gaue.

*A comparison of the Silke flies egges with other egges.*


From *Lybian* egges a mightie (1) bird doth rise,
Scorning both horse and horsemen in the chace,
With Roe-bucks feete, throwing in furious wise,
Dust, grauell, sand and stones at hunters face,
Yet dwels there not beneath the vauted skies,
A greater foole of all the feathred race:
   For if a little bush his head doth hide,
   He thinkes his body cannot be espide.

*1 The Ostrich.*

G

From

1 *The Eagle.* From egges of (1) her whose mate supporteth *Ioue*,
And dares giue combate vnto draggons great,
With whom in vain huge stagges and Lions stroue,
Whose onely sight makes euery bird to sweate,
Whom *Romanes* fed in *Capitole* aboue,
And plac't her Ensigne in the highest seate,
     What else springs out but bloudy birds of praye,
     Sleeping al night, and murdering al the daye?


From egges of famous *Palamedian* foules,
And them that hallow *Diomedes* toomb,
In bodies strange retaining former soules,
VVise, wary, warlike, saging things to come,
VVhose inborne skil our want of witte controules,
Whose timely fore-sight mates our heedlesse doom,
     Comes ought but cranes of most vnseemly shape,
     And diuing Cootes which muddy chanels scrape?


2 *Peacocks.* Yea (2) you whose egges *Hortentius* sometimes sold,
At higher rate then now we prize your sire:
Proud though he be, and spotted al with gold,
Stretching abroad his spangled braue attire,
VVherby, as in a glasse, you do behold,
His courting loue, and longing to aspire:
     VVhat bring ye forth but spectacles of pride,
     VVhose pitchy feete marres al the rest beside?
                         Thrise

Thrife bleffed egges of (1) that renowned dame,    *1 The Pelicane.*
Who bleeds to death, her dead ones to reuiue,
Whome enuious creepers poyfon ouercame,
Whilft fhe fetcht meate to keepe them ftil aliue,
How wel befits her loue that facred *Lamb*,
That heal'd vs all with bleeding iffues fiue?
    Yet hath your fruit this blotte, to ouer-eate,
    And glutton-like to vomit vp their meate.

VVinters (2) *Orpheus* bloudy-breafted (3) Queen,    *2 Robbin-red-*
Sommers fweete folace, nights (4) *Amphion* braue,    *breft.*
Linus (5) delight, *Canaries* clad in greene,    *3 Wrenne.*
All (6) linguifts eke that beg what hart would craue,    *4 Nightin-gale.*
Selling your tongues for euery trifle feene,    *5 Linnet.*
As almonds, nuttes, or what you elfe would haue:    *6 Pies, parrats,*
    Offsprings of egges, what are you but a voice?    *ftares, &c.*
    Angring fometimes your friends with too much
                        (noyfe.

Victorious (7) *Monarch*, fcorning partners all,    *7 The houfe-*
Stowt lions terrour, loue of martial Sire,    *cocke.*
True farmers clocke, nights watchman, feruants call,
Preffing ftil forward, hating to retire,
Conftant in fight, impatient of thral,
Bearing in a little breaft a mighty fire:
    Oh that thou wert as faithful to thy wife,
    As thou art free of courage voice and life!

                    Chafte

Chaste is the Turtle, but yet giuen to hate,
Storkes are officious, yet not voide of guiles,
Hardy are *Haggesses*, but yet giuen to prate,
Faithful are *Dowes*, yet angry otherwhiles,
The whitest swimmer nature e're begate,
Suspition blacke and iealousie defiles:
   Briefely, from egges of euery creature good,
   Sprang nought distainted but this little broode.

As for that (1) egge conceiu'd in idle braine,
Whence flowes (forsooth) that endlesse seed of gold,
The wombe of wealth, the (2) *Nepenthes* of paine,
The horne of health, and what we dearest hold:
I count it but a tale and fable vaine,
By some olde wife, or cousning friar told:
   Supposed true, though time and truth descries,
   That all such workes are but the workes of lies.

*1 Called by Alchimist: Ouum Philosopho-rum, the Philosophers egge.
a A medicine famous in Homer to extinguish all kinds of griefs and paines.*

For when the Sire of truth hath truly saide,
That none can make the couering of his head,
These slender haires, so vile, so soone decaide,
Of so smal worth though nere so finely spread
Shal any witte by humane att and aide,
Transforme base mettals to that essence redde,
   Which buies, not only pearles and precious stones,
   But kingdős, states, & *Monarchs* frō their thrones?

                        Ah

Ah!heau'ns forbid(nay heau'ns forbid it fure,)
That euer Art ſhould more then Nature breede,
Curſe we his worke whoſe fingers moſt impure,
Durſt but to dare the drawing of that ſeede,
Yet when they haue done al they can procure,
And giuen their leaden God a golden weede:
   *Zeuxis* his painted dogge ſhal barke and whine,
   When *Ioue* they turne to *Sol* or *Luna* fine.


*Siſyphian*(1) ſoules, bewitched multipliers,
Surceaſe to pitch this neuer pitched ſtone,
Vaunt not of Natures neſt, nor *Orcus* fires,
Hoping to hatch your addle egge thereon:
Reſtraine in time ſuch ouer-prowd deſires,
Let cre'tures leaue *Creators* workes alone:
   Melt not the golden Sulphur of your hart,
   In following ſtil this fond and fruitleſſe art.

1 *Siſyphus was one of king Æolus ſonnes, delighted in robbing and couſening of his neighbours, wherefore this puniſhment was enioyned him to rowle a*

ſtone continually to the top of a Pyramidall and moſt ſteepe hil, til it reſted there, which was an impoſſible thing to performe, becauſe he could neuer pitch it. *Ouid 3 met.*


Record what once befel great *Aeols*(2) ſonne,
For counterfetting onely but the ſound,
Of heau'nly Canoniers dreadful gunne,
That ſhakes the beams and pillers of this round:
A fiety boult from wrathfull hand did runne,
Driuing falſe forger vnder loweſt ground:
   Where ſtil he liues ſtil wiſhing to be dead,
   Spotted without, within al ſtaind with redde.

2 *Salmoneus, another ſonne to Æolus, who for counterfetting thunder, was turned (as Seruius conceiued) into a Salmon.*

Remem-

Remember eke the Vulture gnawing ftil,
That euer-dying euer-liuing (1) wretch,
VVho ftealingly with an ambitious will,
From *Phœbus* wheeles would vitall fire reach,
Thinking to make by humane art and skill,
His man of clay a liuing breath to fetch:
    Beware in time of like celeftiall rods,
    And feare to touch the onely worke of gods.

*1 Prometheus, fonne of Afa & Iaphet, who enterprifing (as Paracelfus doth) to make man, was tied vppon mount Caucafus in chaines, there to be eaten euerlaftingly by Vultures, and yet neuer to die. Ouid 10. Metam.*

But if you ftill with prowd prefumptuous legges,
VVill needes clime vppe the fiery-fpotted hil,
Pilfring from *Ioue* his Nectar voyde of dregs,
And that immortal meate (2) which none doth fill,
If ye wil needes imbefill thofe faire egges,
VVhich in her child-bedde did their (3) mother kil,
    Yet fay not, that for gifts and vertues rare,
    They do, or may, with thefe my egges compare.

*2 Called Ambrofia.*

*3 Leda, who being gotten with childe by Iupiter in the forme of a fwanne, brought forth two egges, out of the one came Caftor and Clytemneftra, out of the other Pollux and Helena. Hefiodus.*

Thefe, thefe, are they, in dream which Romane fpide
Clos'd in a flender fhell of brittle mould,
Holding within, a white like filuer tride,
VVhofe inward yolke refembleth (5) Ophirs gold,
From out whofe centre fprang the cheefeft pride,
That e're *Latinus,* or his race did hold,
    Exchanging in al countries for the fame, (name.
    Meate, drinke, cloth, coyne, or what you elfe can

*4. Cic. 2. de diu.*

*5 VVhente Salomon fetche gold euerie three yeares, which wifedome would neuer haue permitted him to haue done, if he had knowne (as fome imagine) how to make the Philofophers ftone.*

Here

Here lies the (1)Calx of that renowned shel,
Here flotes that water permanent and cleere,
Here doth the oile of Philosophers dwell,
Stil'd from the golden Fleece that hath no peere:
In midst of whose vnseene and secret cell
Dame Nature sittes,and euery part doth steere,
 Though neither opening shop to euery eie,
 Nor telling (2) *Cæsar* she can multiply.

Al-working mother,Foundresse of this All,
Ten-hundred-thousand-thousand-breasted nurse,
*Dedalian* mouldresse both of great and small,
As large in wealth,as liberall of purse,
Still great with childe,still letting children fall,
Good to the good,not ill vnto the worse,
 VVhat made thee shew thy multiplying pride,
 More in these egges,then all the egges beside?

VVas it,becauſe thou takeſt moſt delight,
To print the greateſt worth in ſmalleſt things?
That they,the leaſt of any ſeede in ſight,
Might clothiers breed to clothe our mightieſt kings?
O witte diuine,O admirable ſpright!
VVorthie the ſongs of him that ſweeteſt ſings:
 Lét it ſuffice that I adore thy name,
 VVhoſe works I ſee,and know not yet the ſame.
                                   But

1 Of which Calx,water,and oyle,you may reade more than enough in Libanius: Epiſt. de ono Philoſophorum, & the troubling Turba Philoſophorum, & the reuerent, D Dee,in Monad.Hierogl.
2 As one or two fooles haue done.
3 A deſcription of Nature.

But damsels, ah : who rustleth in the skie?
Me thinks I heare *Enithean* Ladyes (1) foe,
Blustring in fury from the mountaines hie,
Looke how he raiseth cloudes from dust below,
Harke how for feare the trees do cracke and crie,
Each bud recoyles, the seas turne too and fro:
   O suffer not his breath-bereauing breath,
   To slay your hopes with ouer-timely death.

*1 Boreas, who by
ferce rauished
Orythia King
Ericktheus
daughter, Ouid
6 Metam.*

Therefore assoone as them you gathered haue,
Vpon the whitest papers you can find,
In Boxes cleane your egges full closely saue,
From chilling blast, of deadly nipping winde,
Let not that hoary (2) iry-manteld slaue
So much preuaile, to kill both stocke and kinde:
   Farre be it from a tender Damsels heart,
   On tendrest seedes to shew so hard a part.

*2 Hyems or
winter.*

Yet keepe them not in roomes too hot and close,
Lest heate by stealth encroch it selfe too soone,
And inward matter ripening so dispose,
That spinsters creepe ere winters course be done,
Whilst woods stand bare, & naked ech thing grows,
And *Thisbes* sap for aide be inward runne:
   For as with cold their brooding powre is spilde,
   So ate they then for want of herbage kilde.

*The seedes or
egges of Silke-
flies are to bee
kept neither too
cold, nor any
thing hot.*

                         Th'Arch-

Th'Arch-mafon of this round and glorious bal,
Of creatures created Man the laft,
Not that he thought him therefore worft of all,
(For in his foule part of himfelfe he caft)
But left his wifedome might in queftion fall,
For hauing in his houfe a ftranger plac'e,
  Ere eu'ry thing was made to pleafe and feaft,
  So great a Monarch and fo braue a gueft.

Vnder whofe feete where e're he went abrode
*Vefta*(1) fpread forth a carpet voide of art,
Softer then filke, greener then th'*Emerode*,
Wrought al with flowres, and eu'ry hearb apart,
Ouer him hang'd where e're he made abode,
An azur'd cloth of ftate, which ouerthwart
  Was biaft (as it were) and richly purld,
  With twelue braue fignes & gliftring ftars inurld

*1 The Earth.*

Vppon him then as vaffals eu'ry day
Stowt Lions waited, tameles Panthers eke,
Fierce Eagles, and the wildeft birds of pray,
Huge whales in Seas that mighty carricks wreake,
Serpents and toades: Yea each thing did obey,
Fearing his lawes and ftatutes once to breake:
  Yet wherto feru'd this pompe and honour great,
  If man had wanted due and dayly meate?

H                    Trace

*The seedes or*  Trace you Gods steppes, and til you can attaine
*egs of Silkeflies*  Wherwith to feed your guests when first they shew,
*are not to be*
*hatched till the*  Haste not their hatching, for t'wil prooue a paine,
*Mulberie tree*  Filling your hearts with ruth, your eyes with dew,
*be budded.*  As when th'vntimely lambe on *Sarums* plaine,
 Fallne too too soone from winter-staru'd ewe,
  To pine you see for want of liquid food,
  Which should restore his wants of vitall blood.

*1 The Mulbery*  Attend therefore, when farmers (1) ioy renues
 Her liuely face, and buddeth all in greene,
 For Hyems then, with all his frozen crues,
 Is fully dead, or fled to earths vnseene,
 Corne, cattell, flowers, feare then no heauie newes,
 From Northern coasts, or *Boreas* region keene:
  Birds sing, flies buzze, bees hum, yea al things
  To see the very blush of *Marus* lippe.          (skip

 Let swallowes come, let storkes be seene in skie,
*2 The Nightin-*  Let (2) *Philomela* sing, let (3) *Progne* chide,
*gale.*
*3 The Wrenne.*  Let (4) *Tyry-tiry-leerers* vpward flie,
*4 Larkes.*  Let constant *Cuckoes* cooke on euery side,
 Let mountaine mice abroad in ouert lie,
 Let euery tree thrust foorth her budding pride,
  Yet none can truely warrant winters flight,
  Till she be seene with gemmes and iewels dight.
                                                    O

O peereleſſe tree, whoſe wiſedome is far more
Then any elſe that ſprings from natures wombe:
For though *Pomonaes* (1) daughters budde before,
And forward (2) *Phillis* formoſt euer come,
And *Perſian* (3) fruit yeeldes of her bloſſoms ſtore,
And (4) *Taurus* hotte ſucceedeth (5) *Aries* roome :
  Yet all confeſſe the Mulbery moſt wiſe,
  That neuer breedes  till winter wholly dies.

1 *All kinde of round fruit.*
2 *The Almonde tree.*
3 *Peaches: brought firſt out of Perſia, as Columella, writeth.*
4 *Aprils ſigne.*
5 *March his ſigne.*

Such is her wit : but more her inward might,
For budded newe when *Phœbus* firſt appeares,
She is full leaued e're it grow to night:
With wondrous crackling filling both our eares,
As though one leafe did with another fight,
Striuing who firſt ſhall ſee the heau'nly ſpheares ,
  Euen as a liuely chickin breakes the ſhell,
  Or bleſſed Soules do ſcudde and flie from hell.

Yet witte and ſtrength her pittie doth exceede,
For none ſhe hurts that neere or vnder **grow,**
No not the brire, or any little weede ,
That vpward ſhootes, or groueling creepes below,
Nay more, from heauenly flames each tree is freed
That nigh her dwels, when fearful lightnings glow:
  For vertue which, the Romanes made a law,
  To puniſh them that ſhould her cut or ſaw.

*So writeth Pliny, lib. 10. hiſt. nat.*

Read Pliny,
lib.citato.

I leaue to tell how she doth poison cure,
From adders goare or gall of Lisards got,
VVhat burning blaines she heales and sores impure,
In palat, iawes, and al enflamed throte,
VVhat canckars hard, and wolfes be at her lure,
What Gangrenes stoop that make our toes to rotte:
    Briefly, few griefes from Panders boxe out-flew,
    But here they finde a medcine, old or new.


Her bloud retourn'd to sweete *Thisbean* wine,
Strengthneth the lungs and stomacke ouer-weake,
Her clustred grapes do proue a dish most fine,
VVhose kernels soft do stones in sunder breake:
Her leaues too that conuerted are in time,
Which kings themselues in highest prize do reake:
    Thus giues she meat, and drink, medcine, & cloth,
    To eu'ry one that is not drownd in sloth.

1 So Monardes
calleth it, lib. de
arb. Ind.
2 Leo Afer.

Bragge now no more perle-breeding *Taprobane*,
Of *Cocos* thine, that (1) all-supplying foode,
Vaunt not of Dates thou famous (2) *Africane*,
Though sweete in taste, and swift in making bloud,
Blush *Syrian* grapes, and plums *Armenian*,
*Ebusian* figges, and fruit of *Phillis* good:
    Bad is your best compared with this tree,
    That most delights my little flocke and mee.

                                          But

But wil you know, why this they onely eate?
Why leaues they onely chufe, the fruite forfake?
Why they refufe al choife and fortes of meate,
And hungers heate with onely one difh flake?
Then lift a while, you wonder-feekers great,
Whilft I an anfwere plaine and eafie make:
 Difdaine you not to fee the mighty ods,
  Twixt vertuous worms and finful humane gods.

I thinke that God and nature thought it meete,
The nobleft wormes on nobleft tree to feede:
And therefore they elfe neuer fet their feete
On any tree that beareth fruit or feede:
Others diuine, that they themfelues did weete
No other tree could yeelde their filken threede.
 Iudge learned wittes: But fure a caufe there is,
  VVhy they elfe feede vpon no tree but this.

*Why Silke-
wormes eate on-
ly Mulberie
leaues.*

Ne eate they all, as greedy *Kafers* do,
But leaue the berries to their Soueraigne:
Religioufly forbearing once to bloe
Vpon the fruit, that may their Lord maintaine.
Nay, if thefe leaues (though nothing elfe doth growe
In *Eden* rich their nature to fuftaine)
 Had erft bin giuen for other creatures meate,
  They would haue chufde rather to ftarue then eat.

H 3              In

In that they onely feede vppon one tree,
How iustly do they keepe dame Natures lore?
Who teacheth en'n the bleare-eyde man to see,
That change of meates causeth diseases store:
The gods themselues (if any such there be)
Haue but one(1)meate, one drinke, and neuer more,
 Whereby they liue in health and neuer die,
 For how can one against it selfe replie.

Dualitie of meates was sicknesse spring,
With whom addition meeting by the way,
Begate varietie of euery thing,
Who like a whore in changeable array,
With painted cheekes (as did *Philinus* sing)
And corall lippes, and breasts that naked lay,
 Made vs with vnitie to be at warres,
 And to delight in discords, change, and iarres,

Wherefore assoone as they beginne to creepe,
Like sable-robed Ants, farre smaller tho,
Blacke at the first, like pitch of Syrian deepe,
Yet made in time as white as *Atlas* snow,
Send seruants vp to woods and mountaines steepe,
When Mulb'ry leaues their maiden lippes do shew:
 Feede them therewith (no other soule they craue,
 If morne and eu'n fresh lesage they may haue.)
         The

*(marginal notes:)*
*Why Silke-wormes feed on-ly vpon one meate.*

*1 Called Am-brosia.*
*2 Called Nectar*

*3 Read Plutark*
*4 Sympos, quæst, 1*

The first three weekes the tend'rest leaues are best,
The next, they craue them of a greater size,
The last, the hardest ones they can disgest,
As strength with age increasing doth arise:
After which time all meate they do detest,          *So that they eat*
Lifting vp heads, and feete, and breast to skies,          *not in all aboue*
   Begging as t'were of God and man some shrowde,          *nine weekes.*
   Wherein to worke and hang their golden clowde.

But whilst they feede, let al their foode be drie          *VVhen their*
And pull'd when *Phœbus* face doth brightly shine,          *meate is to bee*
For raine, mist, dewe, and spittings of the skie,          *gathered.*
Haue beene ful of the baine of cattle mine:
Stay therfore, stay, til dayes-vpholder flie,
Fiue stages ful from Easterne *Thetis* line:          *That is to say,*
   Then leaues are free from any poysned seede,          *till the sunne be*
   Which may infect this white and tender breede.          *foue houres high*

Keepe measure too, for though the best you get,          *In what quan-*
Giue not too much nor little of the same,          *titie they are to*
Satiety their stomacks wil vnwhet,          *be dieted.*
Famine againe wil make them leane and lame:
Lend Witte the knife to quarter out their meate,
As neede requires and reason maketh clame:
   Lest belly break, or meagernesse ensewe,
   By giuing more or lesse then was their due.

                     Ne

*Varietie of meate is naught for them.*

Ne cháge their food (tis some haue thought it meet)
For Mulb'ries though they are of double kind,
The blacker ones are yet to them most sweete,
From out their leaues most pleasing sappe they find,
But whé they faile whilst *Scythia* krime 1 doth fleete,
(Turne heau'nly hosts, O turne that cruell wind)
  White Mulb'ry leaues, yea tender Elming bud,
  May for a shift be giuen in steede of foode.

*1 Boreas, the Northwest wind*

*Their table is to be kept cleane.*

Sweepe eu'ry morn ere they fresh vittailes see,
Their papred boord, whereon they take repast,
With bundled Time, or slippes of Rosemary,
Leaue nought thereon that from their bellies past,
No not th'alf-eaten leaues of *Thisbes* tree,
And when their seates perfumed thus thou hast,
  Remooue them back againe with care and heede,
  To former place wherein they erst did feede.

*The sleepe of Silkewormes.*

Oft shalt thou see them carelesse of their meate,
Yea ouer-tane with deepe and heauie sleepe,
Like to that strange and Epidemian sweate,
When deadly slumbers did on *Britons* creepe:
Yet feare thou not, it is but natures feate,
Who nethelesse hath of peerelesse spinsters keepe,
  And makes them thus as dead to lie apart,
  That they may wake and feede with better hearte.
                                        Thrise

Thrife thus they sleep, and thrife they caft their skin,
The latter ftil farre whiter then the reft,
For neuer ate they quiet of mind within,
Til they be cleane of blackneffe difpoffeft,
Whether becaufe they deeme it fhame and finne
To weare the marke of blackifh fiend vnbleft:
  Or that their parents wearing onely white,
  They therefore in that onely would be dight.

*How oft they change their skinnes.*

As they in body and in greatneffe grow,
Diuide them into tribes and colonies,
For though at firft one table and no mo
(Smal though it be) a thoufand wormes fuffice,
Yet afterwards (as proofe wil truly fhow)
When they proceede vnto a greater fize,
  One takes the roome of tenne, and feemes to craue
  A greater fcope and portion for to haue.

*How they are to be diftributed when they grow greater.*

The loft wherein their tables placed be,
Muft neither be too full, nor voide of light,
Two windowes are inough, fuperfluous three,
Plac't in fuch fort that one regard the light
Of *Phœbus* fteeds vprifing as we fee:
And from the other when it drawes to night,
  We may behold them tired as it were,
  And limpiug downe the wefterne *Hemifphere.*

*VVhat manner of roome their table muft ftand in.*

          I         Glafde

Glafde let them be, or linnen-couerd both,
To keepe out fell and blacke (1) *Monopolites*,
The *Myrmedonian* crue, who voide of floth
Do wholy bend their forces, toile, and wittes
To priuate gaine, and therefore are ful wroth
To fee this nation any good befits:
     Working themfelues to death both night & day,
     Not for themfelues, but others to array.

*1 Ants or Emets.*

The greedy imps of her that flue her fonne,
*Pandions* (2) daughter, bloudy harted Queene:
The winged (3) fteedes in *Venus* coach that runne,
Inflam'd with filthy luft and fires vnfeene,
Purfue this flocke, and wifh them al vndone,
Bycaufe they come from parents chafte and cleane:
     O therefore keepe the cafements clofe and faft,
     Left quellers rage your harmeleffe cattle waft.

*2 Wrennes and Robins.*
*3 Sparrowes.*

If alfo carelefneffe haue left a rift,
Or chincke vnftopped in thine aged wall:
Where-through a noyfome mift, or rayny drift,
Or poyfned wind may trouble fpinfters fmall,
Mixe lime and fand, deuife fome prefent fhift
How to repel fuch cruel foe-men al:
     Small is the charge compared with the gaine,
     That fhal furmount thy greateft coft and paine.

I any feeme to haue an amber coate,
And fwell therewith as much as skinne can hold,
Wholy to floth and idleneffe deuote,
Tainting with lothfome gore thē common fold,
Of deadly fickeneffe t'is a certaine note,
VVhofe cure,fith none haue either writte or tolde,
   VVifedom commands to part the dead and ficke,
   Left they infect the faultleffe and the quicke.

*How the ficke are known from the whole, & in what fort to bee vfed.*

Colde fometimes kills them, fometimes ouer-heate,
Raine,oyle,falt,old and wet,and mufty foode,
The fmel of onyons,leekes,garlick,and new wheat,
Shrill founds of trumpets,drums,or cleauing woode:
Yea fome of them are of fuch weakeneffe great,
That whifprings foft of men or falling floud,
   Doth fo their harts and fenfes ouer-wheele,
   That often headlong from the boord they reele.

*Outward caufes of their fickneffe*

Forbeare likewife to touch them more then needes,
Skarre children from them giuen to wantonneffe,
Let not the fruit of thefe your precious feedes,
Die in their hands through too much careleineffe:
VVho toffe and roule and tumble them like weedes
From leafe to leafe in bufie idleneffe,
   Now fquatting them vppon the floore or ground,
   Now fquafhing out their bellies foft and round.

                I 2       Thus

*Signes of their*
*readinesse to*
*worke.*

Thus being kept and fed nine weekes entire,
Surpriz'd with age ere one would thinke them yong,
With what an ardent zeale and hot desire
To recompence thy trauels do they long?
They neither sleepe, nor meate, nor drinke require,
But presse and striue, yea fiercely striue and throng,
　Who first may find some happy bough or broom,
　Whereon to spinne and leaue their amber loome.

*They must*
*scoure them-*
*selues two daies*
*before you set*
*them to worke.*

Then virgins then, with vndefiled hand
Seuer the greatest from the smaller crue,
For al alike in age like ready stand,
Now to begin their rich and oual clue,
(Hauing first paid as Nature doth command,
To bellies-farmer that which was his due)
　For nothing must remaine in body pent,
　Which may defile their sacred monument.

*For that is the*
*best and safest*
*way to loose*
*none of them.*

So being clensde from al that is impure,
Put each within a (1) paper-coffin fine,
Then shal you see what labour they endure,
How farre they passe the weauers craft of line,
VVhat cordage first they make and tackling sure,
To ty thereto their bottom most diuine,
　Rounding themselues ten thousand times & more,
　Yet spinning stil behind and eke before.

　　　　　　　　　　　　　　　　None

None cease to worke:yea rather all contend
Both night and day who shall obtaine the prize
Of working much,and with most speede to end,
Whilst rosie (1) *Titan* nine times doth arise
From purple bedde of his most louing (2) friend,
And eke as oft in (3) *Atlas* vally dies)
 Striuing (a strife not easie here to find)
 In working well,who may exceed their kind.

How they
work not aboue
nine daies.
1 The sunne.
2 Aurora, the
morning.
3 The westerne
sea.

Yea some (O wofull sight)are often found
Striuing,in worke their fellowes to excell,
Lifelesse in midway of their trauerst round,
Nay those that longest here do work and dwell,
Liue but a while,to end their threed renownd,
For I haue seene,and you may see it well,
 After that once their bottoms are begunne,
 Not one suruiues to see the tenth dayes sunne.

Go gallant youths,and die with gallant cheere,
For other bodyes shortly must you haue,
Of higher sort then you enioyed here,
Of worthier state,and of a shape more braue,
Lie but three weekes within your silken beere,
Till Syrian dogge be drownd in westerne waué,
 And in a moment then mongst flying things,
 Receiue not feete alone,but also wings.

How they are
turned into flies
when Dogge
daies end,or
thereabouts.

     I 3   Wings

A description of
the Silkeflies.
1 An exceeding
high hil in Asia
2 Venus Para-
mour, sonne to
Cinara, king of
Cyprus, by his
owne daughter
Myrrha.

Wings whiter then the snow of (1) *Taurus* hie,
Feete fairer then (2) *Adonis* euer had,
Heads, bodies, breasts, and necks of Iuory,
With perfit fauour, and like beautie clad,
Which to commend with some varietie,
And shadow as it were with colour sad,
   Two little duskie feathers shall arise
  From forehead white, to grace your Eben eyes.

VVhen the silke
is to be winded
from the bottom

Then neither shall you see the bottome moue,
Nor any noyse perceiue with quickest eare,
Death rules in all, beneath, in midst, aboue,
Wherefore make haste you damsels voyd of feare,
Shake off delay, as ere you profit loue,
In boxes straite away your bottoms beare,
   Freed from the coffin wherin late they wrought,
  To gaine the golden fleece you so much sought.

In what sort the
silke is to bee
winded.

First pull away the loose and outmost doune,
As huswiues do their ends of knottie towe,
That which lies vpmost is of least renowne,
The finest threed is placed most below:
Threed fitte for kings, vnmeete for euery clowne,
On Natures quill so wound vp rowe by rowe,
   That if thine eye and hand the end can find,
  In water warme thou maist it all vnwind.

                          Three

Three sorts there are, distinct by colours three,
The purest like to (1) their resplendant haire,
Who weeping brothers fal from coursers free,
Their teares were turn'd to yellow amber faire.
The second like (2) her whom impatiencie
Made of a spouse a tree most solitary:
    The last more white, made by the weaker sort,
    Not of so great a price, nor like report.

*How many sorts of silke there be. 1 Phaetusa & Lampetia Phaetons sisters. Ouid 2 Metam. 2 Phillis, Demophoons spouse turned into an Almond tree.*

From out al three, but chiefly from the best,
Are made, not onely robes for priests and kings,
But also many cordial medcins blest,
Curing the wounds that sullen *Saturne* brings,
Which being drunk, how quiet is our rest?
How leaps our hart? how inwardly it springs?
    Speake you sad spirits that did lately feele,
    The hart-breake crush of melancholies wheele.

*The vse of all sorts of silke.*

Nay euen the doune which lies aloft confusde,
Makes Leuant stuffe for country yonkers meete,
Though it of court and cittie be refusde,
And is not worne in any ciuill streete,
But tel me yet, how can (3) he be excusde,
VVho trampled eu'n the best with mired feete,
    And in a moment marr'd al that with pride,
    For making which, tenne thousand spinsters dide?

*3 Diogenes that dogge, who with his dirtie shooes trode downe Platoes silken Quilt (as Laertius writeth) in greater pride then Plato euer vsed it.*

                Now

*The first made bottoms are best to be reserued for seede.*

Now if of these your bottoms you require,
Some to reserue for future race and seede,
Chuse out the eldest, for their forward fire
Makes itiward flye the sooner spring and breede:
Whereas the latter ones haue least desire,
And lesser might to perfit *Venus* deede:
　For why, their pride is dul, and spirits colde,
　Borne in the quarter last of (1) *Iune* olde.

*1 The waining Moone.*

Wind none of them, which you for breede allot,
In watrie bath, nor else in wine, or lye,
Lest outward moisture innly being got,
Surrounding, drownes the little infant-flye,
And cause both strings and secundine to rotte,
So that before it liues it learnes to dye:
　Or if you haue them drenched so for gaine,
　At sunne or fire to dry them take some paine.

*2 That is to say, white paper, for the first writing paper was the inner rinde of a certaine reede or cane, into which Phillira was transformed. Com. Mat. in Mithol. Within 12 daies after the bottoms finished, the silke flies are disclosed.*

Singled, then laye them on a table neate,
Couered al o're with white (2) *Philliraes* skinne,
Stay then againe till *Phœbus* chariot great
In *Oceans* bath hath twelue times washed bin,
And you shal see an admirable seate,
This form'd and yet transformed broode within:
　From which new shapes new bodies do arise,
　And tailes to heads, and worms are turn'd to flies.

Whereat

Whereat to wonder each man may be bold,
When seely worms themselues new fliers made,
Whilst one anothers face they do behold:
Muse how, and when, & where, this forme they had,
How new hornes sprang frō out their forebeads old,
Whence issued wings, which do them ouer-lade:
　　For they recording what they were of late,
　　Dare not yet mount aboue their former state.

*Silke flies feede on nothing but aire.*

As studying thus they stand a day or more,
Offring to feede on nought but onely aire,
Lothing the meate so much desir'd before,
I meane the leaues of *Thisbes* tree most faire:
Disdaining eke to taste of *Nais* store,
To quench the heate that might their harts impaire:
　　At length they know themselues to be aliue,
　　And fal to that for which our wantons striue.

*A day or a little more after disclosing, they couple togither.*

Both long, and longing skud to *Venus* forts,
To stirre vp seed that euer may remaine,
He runnes to her, and she to him resorts,
Each mutually the other entertaine,
Ioynd with such lincks and glue of natures sports.
That coupled stil they rest a day or twaine:
　　Yea oftentimes thrise turnes the welkin round,
　　Ere they are seene vnlocked and vnbound.

*How long they are coupled togither.*

　　　　K　　　　　　　　So

*When they die after discou-pling.*

So hauing left what e're he could impart,
Of spirits, humors, seede, and recrement,
Willing yet further to haue throwne his hart
Into her breaft, to whom he all things ment,
He formoft dies and yeelds to fatal dart :
Ne liues fhe long, but ftrait with forrow fpent,
   (Hauing firft laide the egges fhe did conceiue)
   Of loue and life fhe fhortly takes her leaue.

*Their egges in colour and big-neffe, are likeft of all things to Millet feede, wherewith Par-rachites are fed.*

Smal egges they be, in bigneffe, colour, fhape,
Like to the meate of *Indian* Parrachite,
Leffe farre in view then feed of garden rape,
In number many, yet indefinite :
For when the females womb begins to gape,
And render what the male got ouer night,
   Now more, now fewer feeds dropt from the fame,
   As they were fhort, or longer at their game.

*What number of egges they lay.*

Yet feldome are they than a hundred leffe,
Sometimes two hundred from their loynes do fall,
Round, fmooth, hard-fhelld, and voide of brittlenes,
Whited alike, and yellow yolked all,
Whofe vertues great no man did yet expreffe,
*1 The water or viuer wheron all the Mufes drinke.* Much leffe can I whofe knowledge is fo fmal,
   Though fure I am hence may we find a theame,
   Able to drink vp (1) *Aganippes* ftreame.

O

O keepe them then with moſt attentiue heede,
From *Boreas* blaſt and *Aeols* inſolence,
From menſtruous blaſts & breathing keep thē freed,
Auoide likewiſe the mil-dewes influence,
Pray heau'nly *Monarch* fot to bleſſe your ſeede,
Helping their weakneſſe with his prouidence:
  So may your milk-white ſpinſters worke amaine,
  When *Morus* lippes ſhal bud and bluſh againe.

*How, the egges are to be preſerued.*

And (1) thou whoſe trade is beſt and oldeſt too,
Steward of all that euer Nature gaue,
VVithout whoſe help what can our rulers doo,
Though gods on earth appareld wondrous braue?
Behold thy helping hand faire virgins wooe,
Yea nature bids, and reaſon eake doth craue
  Thy cunning, now theſe little worms to nurſe,
  VVhich ſhal in time with gold fill full thy purſe.

*1 An exhortation to all Farmers and Huſbandmen to plant Mulberries.*

In ſteed of fruitles elms and ſallowes gray,
Of brittle Aſh, and poyſon-breathing vgh,
Plant Mulb'ry trees nigh euery path and way,
Shortly from whence more profit ſhal enſue,
Then from (2) th'Heſperian wood, or orchards gay,
On euery tree where golden apples grew:
  For what is ſilke but eu'n a Quinteſſence,
  Made without hands beyond al humane ſenſe?

*2 Made and planted by Ægle, Arethuſa, & Hyperethuſa, King Atlas daughter.*

A commendati-
on of this silke,
with that which
commeth from
the Offereaus,
as also with
that which is
made by the In-
dian wormes.

A quinteffence? nay wel it may be call'd,
A deathleffe tincture, fent vs from the skies,
Whose colour ftands, whose gloffe is ne're appalld,
Whose Mulbr'y-fent and fauour neuer dies,
Yea when to time all natures elfe be thralld,
And euery thing Fate to corruption ties:
   This onely fcornes within her lifts to dwell,
   Bettring with age, in colour, gloffe, and fmel.

2 Of thefe Of-
ferians or Lords
of the wood,
read Bonfin. lib.
1 Decad 1.
Hung. Hift.
2 Aurelianus
furnamed the
Liberall, liuing
274.yeares af-
ter Chrift, in

So doth not yours (you (1) Lordings of the woode)
Growing like webbs vppon the long-haird graffe,
Along the (2) Offerian bancks of Scithyan floud,
Which into Cafpian wombe doth headlong paffe.
No, no: Although that filke be ftrong and good
In outward fhew, and highly prized was,
   When bounteous Cæfar ruled citties prime,
   Yet foone it fades, and yeelds to rotte in time.

whofe time a pound weight of filke was fold for the like weight in fine gold. Vopifcus.

3 Paufanias
bookes.
4 The Dor-
beetle.
5 The Spider.
6 The Reede or
cane.
7 The hie oakes.

If (3) bookes be true, there is an Indian worme,
As bigge as (4) he that robbs the Eagles neft,
Shap't like (5) Arachne that doth tinfels forme,
And nets, and lawnes, and fhadowes of the beft,
Fed with (6) her locks, who yeelding ftands in ftorm,
VVhen (7) woods-furueyours lye on earth oppreft)
   From out whofe belly, broke with furfetting,
   VVhole clews of filk fcarfe half concocted, fpring.
                                                    Yet

Yet that compar'd with this is nought so fine,
Ne ought so sweetely fum'd with daintie sent,
Nor of like durance, not like powre diuine:
Mirth to restore, when spirits all are spent,
If it be steept in sweet *Pomanaes* (1) wine,
Till colour fade, and substance do relent:
  Nay, nay, no silke must make that (2) Antidote,
  Saue onely which from spinsters mine is got.

1 The *goddesse*
of apples.
2 Called Con-
fectio Alkermes
a most singular
Electuarie a-
gainst Melan-
cholie, if it be
rightly made.
3 Io. Fernelius.
lib.7 qui est de
composured.

Whereof, if thou a pound in weight shalt take
Vnstaind at all (as *Amiens* (3) floure doth write)
And with the iuce of Rose and pippins make
A strong infusion of some day and night,
Adding some graines of muske and Ambres flake,
And seething all to hony-substance right:
  O what a Balme is made to cheere the heart,
  If pearle, and gold, and spices beare a part?

What neede I count how many winders liue,
How many twisters eke, and weauers thriue
Vppon this trade? which foode doth daily giue
To such as else with famine needes must striue:
What multitudes of poore doth it relieue,
That otherwise could scarce be kept aliue?
  Say Spaniard proude, & tel Italian youth,
  Whether I faine, or write the words of truth.

K 3                    Not

Not euer were your princes clad so braue,
Not euer were your wiues deckt as they be,

*1 Heliogabalus,*
*for so writeth*
*Lampridius.*

Much lesse was silk then worne of euerie slaue,
And artists, sprung from base and low degree,
**That** (1)rioter whose belly diggd his graue,
Clothd all in silke, the Romanes first did see:
   Before whose time silke wou'n on linnen threed,
   Was thought braue stuffe for any Princes weed.

*VVhen the*
*soede of silke-*
*wormes was first*
*brought into*
**Europe.**

*So Polidor vir-*
*gil writeth out*
*of Procopius,*
*saying that this*
*happened 555.*
*yeares after*
*Christ, lib.3*
*cap.6.de ret. in-*
*uent,*
*2 A citie of east*
*India.*

But afterwardes, when holy Palmers twaine
From out (2)*Serinda* brought these worms of fame,
And plauted Mulb'ry plants on hill and plaine,
Wherewith to fatte and foster vppe the same:
How rich waxt *Italy?* how braue was *Spaine?*
In Sattin fine, how braggd each man of name?
   Yea, euery clowne, that euen as now, so then,
   Habites did scarce discerne the states of men.

Vp Britaine blouds, rise hearts of English race,
Why should your clothes be courser then the rest?
Whose feature tall, and high aspiring face,
Aime at great things, and challenge eu'n the best.
Begge countrymen no more in sackcloth base,
Being by me of such a trade possest:
   That shall enrich your selues and children more,
   Then ere it did *Naples* or *Spaine* before.

No

No man ſo poore but he may Mulb'ries plant,
No plant ſo ſmal but wil a ſilke-worme ſeede,
No worme ſo little(vnleſſe care do want)
But from it ſelfe wil make a clew of threede,
Ech clew weighs down,rather with more then ſcant,
A penny weight,from out whoſe hidden ſeede,
　(After the winged wormes conception)
　A hundred ſpinſters iſſue forth of one,

*How taſie and chargeleſſe a thing it is to keep ſilkworms.*

*What ouerplus there is in profite by keeping them.*

Diuine we hence,or rather reckon right,
What vſury and proffit doth ariſe,
By keeping well theſe little creatures white,
Worthy the care of euery nation wiſe,
That in their owne or publique wealth delight.
And raſhly wil not things ſo rare deſpiſe:
　Yea ſure,in time they wil ſuch profit bring,
　As ſhall enrich both people,prieſt,and king.

Concerning pleaſure : who doth not admire,
And in admiring,ſmiles not in his hart.
To ſee an egge a worme,a worme a flier,
Hauing firſt ſhewd her rare and peereleſſe art,
In making that which princes doth attire,
And is the baſe of euery famous Mart?
　And then to ſee the flie caſt ſo much ſeede,
　As doth, or may, an hundred ſpinſters breede.

*How great pleaſure there is in keeping them, both to the eies, eares,noſe,and hands.*

Againe

Againe to view vppon one birchen shredde,
Some hundred Clewes to hang like clustred peares,
Those greene, these pale, and others somewhat red,
Some like the locks hanging downe *Phœbus* eares:
And then, how Nature when each worme is dead,
To better state in tenne dayes space it reares:
    Who sees all this, and tickleth not in minde?
    To marke the choyse and pleasures in each kinde.


Eye but their egges, (as Grecians terme them well)
And with a penne-knife keene diuide them quite,
Behold their white, their yolke, their skin, and shel,
Distinct in colour, substance, forme, and sight:
And if thy bodies watchmen do not swell,
And cause thee both to leape and laugh outright,
    Thinke God and nature hath that eye denied,
    By which thou shouldst frō brutish beasts be tried.


When they are worms, mark how they color chāge,
From blacke to browne, from browne to sorrel bay,
From bay to dunne, from dunne to duskie strange,
Then to an yron, then to a dapple gray,
And how each morne in habites new they range,
Till at the length they see that happy day,
    When (like their Sires and heau'nly angels blest)
    Of pure and milk-white stoles they are possest.
                            Large

Lay then thine eare and listen but a while,
Whilst each their foode from leafage fresh receaues,
Trie if thou canst hold in an outward smile,
When both thine eare and phantasie conceaues,
Not worms to feed, but showrings to distil.
In whispring sort vpon the tatling leaues:
    For such a kind of muttring haue I heard, (teard.
    Whilst herbage greene with vnseene teeth they

When afterward with needle pointed tongue,
The Flies haue bor'd a passage through their clewes,
Obserue their gate and steerage al along,
Their salutations, couplings, and *Adieus*:
Heare eke their hurring aud their churring song,
When hot *Priapus* loue and lust renewes,
    And tel me if thou heardst, or e're didst eye,
    Like sport amongst all winged troupes that flye.

Tis likewise sport to heare how man and maide,
Whilst winding, twisting, and in weauing, thay
Now laugh, now chide, now scan what others saide,
Now sing a Carrol, now a louers lay,
Now make the trembling beames to cry for aide,
On clattring treddles whilst they roughly play:
    Resembling in their rising and their falls,
    A musicke strange of new found *Claricalls*.
              L       The

The smel likewise of silken wool that's new,
To heart and head what comfort doth it bring,
Whilst we it wind and tooze from oual clew?
Resembling much in prime of fragrant spring,
When wild-rose buds in greene and pleasant hue,
Perfume the ayre, and vpward sents do fling,
    Well pleasing sents, neither too sowre nor sweete,
    But rightly mixt, and of a temper meete.


As for the hand, looke how a louer wise
Delighteth more to touch *Astarte* slick
Then *Hecuba*, whose eye-browes hide her eies,
Whose wrinckled lippes in kissing seeme to prick,
Vpon whose palmes such warts and hurtells rise,
As may in poulder grate a nutmegge thick:
    So ioy our hands in silke, and seeme ful loth
    To handle ought but silke and silken cloth.


Such are the pleasures, and farre more then these,
Which head, and hart, eies, eares, and nose, and hands,
Take, or may take, in learning at their ease,
The dieting of these my spinning bands,
VVhoSe silken threede shal more then counterpeise,
Paine, cost, and charge, what euer it vs stands,
    So that if gaine or pleasure can perswade,
    Go we, let vs learne the silken staplers trade.

                                   But

But lift, me thinkes I heare *Amyntas* fayne,
That fhepheards skill wil foone be quite vndone,
Behold faire *Phillis* fcuddeth from the plaine,
Leauing her flocks at randon for to runne,
Lo *Lidian* clothier breaks his loomes in twaine,
And thoufand fpinfters burne their woollen fpunne:
  Ah!ceafe your rage,thefe fpinfters hurt you nought
  But wil encreafe you more then ere you thought.

*Keeping of filke-
wormes hindreth
not the keeping
of fheepe nor,
Sheepheards.*

For carde an ounce of filke with ten of wooll,
How fine,how ftróg,how ftrange a yarne doth rife?
Make trial once,and hauing feene at ful,
Your new found ftuffe,chaffred at higheft prize,
Then blame your idle heads and fenfes dull,
Truft not conceit,but credite moft your eyes:
  Laughing as much,or more, thē ere you mourn'd,
  When feare you fee to ioy and vantage turnd.

Laugh now (faire *Mira*)with thy Virginswhite,
For why your egges committed to my care,
Are growne fo much in bigneffe,worth,and fight,
That Kings and Queens to keep them wil not fpare,
Yea Queen of Queenes,for vertue,witte,and might,
Perhaps wil hatch them twixt thofe hillocks rare,
  Where al the *Graces* feede and *Sifters* nine,
  Who euer loue, and grace both thee and thine.
FINIS.